Powerful women in history

Phillips Tahuer

Ediciones Afrodita

Contents

Introduction

Throughout human history, numerous women have left an indelible mark in diverse fields, from science and politics to art and human rights. Yet for centuries, prevailing narratives have tended to marginalize or even omit women's contributions, especially at times when patriarchal structures limited their opportunities and recognition. This book, Powerful Women in History, seeks to highlight the lives and legacies of twenty-five exceptional women, whose actions and achievements changed the course of their disciplines, inspired millions, and will continue to influence our society today.

What defines a powerful woman? The answer is neither simple nor one-dimensional. In these pages we will explore a variety of forms of power: the intellectual power of Marie Curie, who transformed the field of physics and chemistry with her radioactive research; the political power of Indira Gandhi, who led India during one of its most complex and challenging periods; and the transformative power of figures like Harriet Tubman, whose unwavering courage saved the lives of hundreds of slaves and raised social awareness about slavery in the United States.

The word "powerful" encompasses both visible and silent impact. In this sense, some of the women featured here directly challenged norms and structures, while others worked within, often facing reprisals, isolation, and personal sacrifice to remain true to their ideals. Women like Simone de Beauvoir not only transformed feminist thought but opened a new path of introspection and understanding about gender equality, redefining relations between the sexes. Meanwhile, figures like Wangari Maathai changed the course of history by connecting environmental conservation with human rights, something unthinkable in her time and still at the forefront of social and ecological action today.

Each woman in this book comes from a distinct context: cultures, times, and obstacles that varied widely. Florence Nightingale, one of the pioneers of modern nursing, redefined health care in an era when the rights and dignity of the sick were ignored. In the contemporary era, women like Malala Yousafzai remind us that the fight for education and justice continues, even in contexts of extreme adversity.

The power of these women also lies in their differences. Through their biographies, we will see how figures like Cleopatra and Hatshepsut used their ingenuity and determination to lead Egypt and protect it from internal and external threats; women like Frida Kahlo, who channeled her personal pain into her art, turning it into a tool for deep and authentic expression of the human experience; and figures like Emmeline Pankhurst, who organized an unprecedented fight for women's suffrage, winning fundamental rights for future generations of women.

The intention of this book is not only to recount their lives, but also to offer a reflection on their impact on the modern world. How is a woman's legacy measured in the context of her time and the centuries that followed? How are her contributions understood in a world where, despite advances, women continue to face inequalities? Each story we will read in these pages invites us to remember that power can come in many forms and that the actions of these women not only generated changes in their own lives, but also laid the groundwork for the social, scientific and political advances of which we are beneficiaries today.

This book is a tribute to the strength, intelligence, creativity and courage of these twenty-five women. Their biographies represent a celebration of their achievements and an opportunity to remember that the power of one person to change the world has no gender or limitations. With these stories, we want each reader to find inspiration and feel called to pursue their own dreams and fight for a more equitable world. Because, as these women teach us, the power of determination, knowledge and integrity has the potential to transform entire societies.

One of the most evident themes when exploring the lives of these women is their courage in challenging the stereotypes of their time. In this way, Amelia Earhart challenged the perception that flight was exclusively male territory, proving that daring and talent for aviation were not dependent on gender.

Many of these women, by confronting gender barriers and limiting expectations, demonstrated that leadership ability, ingenuity and emotional strength are inherent qualities of being human and not dependent on any social construct. Through their acts of defiance and perseverance, they made entire generations reconsider what is possible and who is allowed to achieve success in any field.

Each of the women in this book has left a legacy that remains relevant today. At a time when gender equality is still an unmet goal in many parts of the world, stories like these women remind us how far we have come and what remains to be achieved. Their lives show that history is not just a series of events, but a fabric of individual struggles, challenges and achievements that together chart the progress of humanity.

People like Ada Lovelace, whose pioneering work on algorithms for Charles Babbage's analytical engine laid the foundations for computer programming, remind us that the field of science and technology is due to the contributions of many women who, in their time, were not recognized as they should have been. In the modern world, where technology plays a fundamental role in our lives, Lovelace's work is a clear reminder that innovation has no gender and that the recognition of women in STEM (science, technology, engineering and mathematics) is essential to inspire future generations.

This book not only tells the story of these women as something distant, but also aims to inspire and empower new generations. Each biography contains moments that invite reflection and remind us that the fight for equality and justice is an ongoing journey. Through these stories, we hope that readers will find not only admiration, but also a stimulus to act in their own lives, with the same conviction, courage and ethics as those who preceded them.

It is important to remember that power is not measured only in terms of visible achievements. In many cases, it is determination and resilience, maintained over years and in difficult circumstances, that define a powerful person. The strength of these women lies not only in the successes they achieved, but in their ability to endure, to imagine a different world, and to act decisively and with hope in moments when others would have lost faith. From Florence Nightingale's dedication to caring for the wounded to the visionary leadership of women like Angela Merkel, each of these stories shows us the capacity of a single human being to make a difference.

Powerful Women in History is a journey through centuries of advances achieved by women who not only excelled in their areas of expertise, but also redefined what power means. Their stories are a reminder that every person can contribute to change and that the limits imposed by society can be overcome with courage and determination.

In these pages, we hope to provide a tribute and a platform for the voices of those who, against all expectations, managed to make a difference. May the lives of these women serve not only as a tribute to their legacy, but as a lasting inspiration to all those who dream of a better future and are willing to fight to build it. Because by remembering their stories, we also remember the power that lies within each of us to create a more just, equitable world full of opportunities for all.

1. Cleopatra VII

One of the most fascinating female figures in ancient history, Cleopatra was born in 69 BC in Alexandria to Ptolemy XII. Her life and reign (51-30 BC) were marked by a complex web of political alliances and an intense struggle to preserve Egypt's independence in the face of the expanding Roman Empire.

Cleopatra excelled in politics for her cunning and ability to form strategic alliances, especially with high-ranking Roman leaders such as Julius Caesar and Mark Antony. As queen, she understood that the balance of power in the Mediterranean depended largely on Rome. To protect her country, Cleopatra used both her intellect and her charisma.

She spoke several languages—including Greek, Egyptian, and possibly Hebrew and Arabic—and was the only one of her dynasties to identify as Egyptian at a time when Hellenistic culture dominated the region. This ability to adapt culturally allowed her to better connect with the Egyptian people and reinforce her legitimacy as a ruler.

Her first major alliance was with Julius Caesar, whom she welcomed to Egypt in the context of a civil war with her brother Ptolemy XIII, with whom she co-ruled. Cleopatra, eager to control the kingdom, used her intelligence and charm to win Caesar's support, thus securing the throne of Egypt. From this union was born her son Caesarion, who symbolized the alliance between Rome and Egypt. Although the relationship with Caesar ended abruptly with his assassination in 44 BC, Cleopatra maintained her power and independence through her shrewd politics.

Later, Cleopatra established another powerful alliance with Mark Antony, one of the most influential leaders in Rome after Caesar's death. With him she shared a personal and political relationship that transformed into a pact of love and power. Cleopatra financed Antony's military campaigns, and together they attempted to consolidate an eastern empire that could have changed the balance of power in Rome. The famous Battle of Actium in 31 BC marked the end of this dream, and after her defeat by Octavian (future Emperor Augustus), Cleopatra chose to take her own life in 30 BC, thus avoiding being displayed as a prisoner in Rome.

Cleopatra's legacy has endured throughout the centuries, not only as a queen of great intelligence and charm, but as a complex cultural figure. Her ability to play the political game, her defense of Egypt, and her determination made her a symbol of female power.

Her death marked the end of Egypt's independence, which was annexed to Rome, but her cultural influence continued. Throughout the centuries, Cleopatra has been a source of inspiration in literature, art, and film. She has been depicted as both a seductive and exotic figure and a fierce strategist and wise ruler, showing the duality of her life and personality.

Cleopatra VII left a legacy as one of the most powerful women in history, combining charisma, intelligence and calculated politics to maintain her kingdom against Roman might. Her story is a reminder of female complexity and resilience in a world ruled overwhelmingly by men, and remains an inspiration to those interested in power, diplomacy and ancient history.

2. Joan of Arc

She was an iconic figure in French history and is remembered as one of the country's greatest heroines. She was born in 1412 in Domrémy, a small village in northeastern France, at a time of war and crisis for the kingdom. The Hundred Years' War between France and England was at a critical point, and much of French territory was occupied by the English and their Burgundian allies. Against this backdrop, Joan emerged as a military leader inspired by her religious visions, and at the age of 17 she became a key player in the French victory in the war.

Joan of Arc excelled in the military arena, not so much for her skill as a strategist, but for her charisma and ability to inspire troops at a time when the country was divided and the army demoralized. From a young age, Joan claimed to hear voices of saints and angels calling her to save France and help Charles VII regain his throne. Guided by these visions, she approached Charles' court and asked for permission to lead her troops against the English. Despite initial misgivings, Charles granted her permission, and Joan, clad in armor, took up arms and won the trust of the French army and people.

Joan of Arc distinguished herself from the Battle of Orleans, which was her first major victory in 1429 and a turning point in the war. In this battle, she led the attack with unwavering bravery, inspiring the troops and achieving a decisive victory that broke the English siege and restored the morale of the French. Following the victory at Orleans, Joan accompanied Charles VII on a series of campaigns that cleared the way for him to be crowned in the cathedral of Reims. This symbolic act reaffirmed Charles' right to the throne and gave new impetus to the French side.

However, Joan's career was short and tragic. In 1430, while defending the city of Compiègne, she was captured by the Burgundians, allies of the English, who handed her over to their enemies. She was put on trial in Rouen, conducted by a pro-English ecclesiastical court that accused her of heresy, apostasy, and wearing men's clothing, among other charges. During the trial, Joan defended her faith and her visions staunchly but was ultimately convicted and sentenced to death at the stake. On May 30, 1431, at the age of 19, Joan of Arc was executed in the town square of Rouen.

Her legacy is vast and complex. Death at the stake was seen as an act of martyrdom and sparked outrage among many in France, making her a symbol of resistance and patriotism. In 1456, a second court reviewed the trial and acquitted her of all charges, restoring her honor. Finally, in 1920, the Catholic Church canonized her, making her a saint and reaffirming her place as one of France's most important religious figures.

Joan of Arc left a legacy of bravery and sacrifice that transcended medieval times and remains relevant today. She was a young woman who defied gender expectations, defended her country with conviction, and managed to inspire a nation in a time of hopelessness. Her life and sacrifice continue to be a symbol of faith, patriotism, and courage, and she is remembered as a model of leadership and dedication to a higher purpose.

3. Elizabeth I of England

Born on September 7, 1533, in Greenwich, London, Elizabeth was the daughter of Henry VIII and Anne Boleyn, and her life was marked by the religious conflicts, political disputes, and dynastic changes that shook England in the 16th century. Elizabeth ascended to the throne in 1558 and reigned until her death in 1603, a period of almost 45 years that is known as the Elizabethan Age, a time of cultural flourishing, exploration, and consolidation of England as a great European power.

Elizabeth I excelled in the political and diplomatic arena in an extremely volatile era. Her life, from the beginning, was surrounded by danger: her mother was executed when Elizabeth was only two years old and left without a mother and without the right of succession, her position at court was unstable. However, her intelligence and education prepared her for a life as a ruler. Upon taking the throne, she faced an England divided between Catholics and Protestants and with a weak economy. Elizabeth demonstrated her diplomatic skill by carefully navigating this religious context: she developed a moderate version of Protestantism, known as the Anglican Church, which allowed her to consolidate control and reduce internal tensions.

Her political skill was also reflected in her decision to remain single. Known as "The Virgin Queen," Elizabeth used her single status to establish alliances with other European powers, using the possibility of marriage as a diplomatic tool without losing the political control that could have been compromised by marriage.

During her reign, Elizabeth fostered cultural growth and supported playwrights such as William Shakespeare and Christopher Marlowe, who transformed English theater and laid the foundations for modern literature. Elizabeth's time also saw the flourishing of music, art, and poetry. Her court became a cultural and political center, and this artistic prosperity spread throughout the country.

Furthermore, Elizabeth supported maritime exploration, which led to the expansion of trade and the beginning of English colonialism. Under her patronage, explorers such as Sir Francis Drake and Sir Walter Raleigh undertook expeditions that challenged the Spanish navy and opened new trade routes. These explorers laid the first foundations of the British Empire, and their victory over the Spanish Armada in 1588 marked the beginning of English naval supremacy.

One of Elizabeth's greatest achievements was the defense of England against the Spanish Armada, sent by Philip II of Spain in 1588. In a war that seemed unfavorable to the English, Elizabeth inspired her troops with her famous speech at Tilbury, where she proclaimed that she had "the heart and stomach of a king." The defeat of the Spanish Armada was a decisive moment that consolidated England as a maritime power and marked the decline of Spanish power.

The legacy of Elizabeth I is immense. Her reign strengthened English national identity and was a period of unprecedented political stability and cultural splendor. By leaving a strong, centralized, and prosperous kingdom, Elizabeth laid the foundations for the future British Empire. Her moderate religious policies, her ability to avoid internal conflict, and her ability to wield power without the intervention of a husband made her a model of female leadership in a male-dominated age.

Her life and reign have been celebrated in history and popular culture as an example of political intelligence and determination. Elizabeth I left behind a more united, powerful, and culturally rich England, and her reign is still remembered as a golden age.

4. Sor Juana Inés de la Cruz

Born as Juana Inés de Asbaje y Ramírez de Santillana on November 12, 1651, in San Miguel Nepantla, Mexico, she is one of the most prominent figures of the Golden Age of Spanish literature. Her intelligence, literary talent and passion for knowledge led her to be one of the most influential writers, poets and thinkers of her time. In addition to her achievements in literature, her life represents a fight for women's rights to education and freedom of expression, which makes her an inspiring figure ahead of her time.

Sor Juana excelled in the literary field, especially in poetry, theater and prose. From a very young age, she showed an insatiable thirst for knowledge, learning to read at the age of three and quickly standing out for her ability in poetry. In her teens, she was introduced to the viceregal court of New Spain (now Mexico), where she impressed the intellectuals of the time with her knowledge and mental acuity. She soon decided to join the religious life at the convent of San Jerónimo in Mexico City, where she found a space to devote herself to her studies and her writing.

Sor Juana wrote a wide variety of works, including sonnets, redondillas, carols, essays and plays. Her poetry, which addresses love, philosophy and theology, stands out for its depth and its elegant and precise language. She was also characterized by including in her works a sharp criticism of the injustices of her time, especially regarding the position and limitations of women in colonial society. Her best-known poem, "Hombres necios que acusáis" (Foolish Men Who Accuse), is a denunciation of the hypocrisy and double standards of men towards women.

In addition to her literary talent, Sor Juana was one of the first women in America to openly defend women's right to education and access to knowledge. Her most famous work in this regard is the "Response to Sor Filotea de la Cruz," a letter in which she defends her right to study and write, challenging the ecclesiastical authority that had asked her to abandon her intellectual dedication. In this letter, Sor Juana argues that the desire for knowledge is a virtue, and that learning is a way to honor God, regardless of gender.

Sor Juana Inés de la Cruz's legacy has endured for centuries, becoming an inspiration for feminist movements and a symbol of the fight for equality and women's rights. Her work has been studied and admired throughout the world, and she is considered one of the greatest poets in Spanish literature. Sor Juana has become a symbol of resistance and the defense of intellectual freedom and expression, and her life and work have inspired writers, scholars, and activists throughout the years.

Today, Sor Juana is remembered as the "Tenth Muse" and her legacy lives on, reminding us of the importance of access to knowledge, freedom of thought, and women's right to education and culture. Her voice continues to be relevant and admired throughout the Spanish-speaking world and beyond, consolidating her as one of the most important literary and cultural figures in Latin America and the world.

5. Catherine the Great

Born Sophia Frederica Augusta of Anhalt-Zerbst on 2 May 1729 in Stettin (then part of Prussia, now Poland), she was Empress of Russia from 1762 until her death in 1796. During her 34-year reign, Catherine transformed Russia into a major European power, promoting the modernization of its institutions, territorial expansion, and the flourishing of arts and culture. Her forward-thinking vision and strong leadership made her one of the most prominent figures in Russian and European history.

Catherine excelled in the political and reformist arena, marking a new era of progress and modernization in Russia. Inspired by the ideals of the Enlightenment, Catherine undertook a series of reforms to improve government, the economy, and education in Russia. In 1767, she commissioned a code of laws that sought to reform justice and protect the rights of subjects, although the reforms were never fully implemented due to the complexity of the feudal system and resistance from the nobility. Still, Catherine promoted trade, encouraged the development of industry, and supported the creation of educational institutions.

In her rule, Catherine implemented a series of reforms to reorganize Russia's regional administration, increasing government efficiency and control over the provinces. She also supported the education of women and encouraged the creation of hospitals, orphanages, and schools, to modernize Russian society and bring it closer to European ideas.

One of Catherine's greatest achievements was the territorial expansion of Russia, which expanded considerably during her reign. Catherine waged successful wars against the Ottoman Empire and expanded Russian control in the Black Sea region, establishing Russian access to important trade routes. In the West, under her rule, Russia became one of the powers that partitioned Poland and acquired vast portions of its territory through the Partitions of Poland (1772, 1793, and 1795), thereby consolidating its influence in Eastern Europe.

Her military policy was also crucial in establishing Russia as a major European power. Under her leadership, the Russian army grew stronger and won victories that reaffirmed Russia's presence and power on the European scene.

Catherine the Great was a great promoter of culture and the arts. She was an avid reader and corresponded with leading philosophers and intellectuals of the time, such as Voltaire and Diderot, which connected her to the circles of the European Enlightenment. Her interest in education and culture led her to find the famous Hermitage in St. Petersburg, which began as her personal art collection and grew into one of the largest and most prestigious museums in the world.

She also supported literature, theatre and the sciences, contributing to Russia's intellectual growth and fostering the creation of a national culture. Catherine also wrote several plays and essays reflecting politics, philosophy and education.

Catherine is remembered as one of the most influential monarchs in Russian history, admired for her steadiness, vision and dedication to the country's greatness. Her reign transformed Russia, and she remains one of the most powerful and complex figures in European history. Through her policies and her love of the arts and the Enlightenment, Catherine left a cultural and historical legacy that continues to inspire Russia and the world.

6. Ada Lovelace

Born Augusta Ada Byron on December 10, 1815, in London, she was a British mathematician and writer, recognized for her innovative contributions to computing and considered the first programmer in history. Daughter of the famous Lord Byron and the mathematician Anne Isabella Milbanke, Ada showed a great aptitude for mathematics and science from an early age, greatly influenced by her mother, who encouraged her to cultivate a rational and analytical education. Despite having lived in the 19th century, in an environment that limited opportunities for women in the scientific field, Ada Lovelace broke barriers and dedicated herself to the study of mathematics, achieving a notable collaboration with the mathematician Charles Babbage in his invention of the analytical engine.

Ada excelled in the field of mathematics and computing. Her talent and curiosity led her to meet Charles Babbage, a pioneer of computing, with whom she developed a friendship and a professional collaboration. Babbage was working on his "Analytical Engine," an advanced mechanical computer design that, while never fully completed, was a conceptual precursor to modern computers. Ada became deeply interested in Babbage's project and was invited to translate into English a paper written by Italian mathematician Luigi Federico Menabrea, which described the workings of the Analytical Engine.

However, her most important contribution went far beyond mere translation. Ada added a set of notes of her own, known as "Ada's Notes," which tripled the size of the original paper. These notes contain analysis and reflections that anticipate many ideas about modern computing. Lovelace worked out a detailed algorithm that would allow the Analytical Engine to calculate Bernoulli numbers, becoming the first algorithm designed to be processed by a machine. This achievement is what today consecrates her as the first programmer in history.

One of Ada's most visionary contributions was her realization that computers could go beyond mathematical calculations and could manipulate symbols and process information in a wide variety of ways. This was revolutionary in an era when machines were seen solely as calculation tools. Ada imagined that one day machines could create music, compose pieces of art, and even perform complex tasks involving data processing and logical reasoning. Her famous statement that the Analytical Engine could eventually "act on things other than numbers" is seen today as a key point in the history of computing.

Ada Lovelace died young, at 36, from uterine cancer, and during her lifetime her work went largely unnoticed. It was not until the 20th century, with the development of modern computing, that her contributions began to receive the recognition they deserved. Today, her work is considered one of the first visions of what computer science would become. In 1979, the United States Department of Defense created the programming language "Ada" in her honor, recognizing her pioneering role in the history of programming.

Ada Lovelace's legacy extends beyond her technical contributions. Her life and work symbolize the importance of women's inclusion in the scientific and technological realm. Her innovative thinking and ability to imagine the future of machines as universal tools have inspired generations of scientists, mathematicians, and programmers. Lovelace remains an inspiring figure and is celebrated as one of the brightest and most pioneering minds in science and technology, a model of intelligence and creativity in a field historically dominated by men.

7. Harriet Tubman

Born in March 1822 in Dorchester County, Maryland, she was a leading figure in the fight to abolish slavery in the United States and one of the most celebrated activists of the 19th century. Of African descent and born into slavery, Tubman risked her life on multiple occasions to free dozens of enslaved people through the network of clandestine routes known as the "Underground Railroad." Her bravery and determination made her a symbol of freedom and resistance in American history. In addition to her role as an abolitionist, Tubman was a nurse, spy, and activist for civil and women's rights, leaving a legacy of courage and compassion.

Harriet Tubman excelled in the field of abolitionist activism. In 1849, she escaped slavery and headed north, achieving her freedom. However, her own freedom was not enough for her, and she repeatedly returned to the South, braving danger and risking her life to help other enslaved people escape. During the 1850s and 1860s, Tubman made at least thirteen missions to free approximately seventy people, including family and friends, using a secret network of routes, safe houses, and willing helpers in what became known as the Underground Railroad.

Tubman was known as "Moses" among the enslaved, in reference to the biblical leader who freed his people from slavery. Displaying extraordinary intelligence and determination, she never lost a single person in her care. Thanks to her ability to avoid capture and her bravery in the face of danger, she became one of the most recognized and respected activists of the era.

When the American Civil War began in 1861, Tubman saw an opportunity to contribute to the Union cause and the liberation of the enslaved in the South. She worked as a nurse in field hospitals, treating wounded soldiers and enslaved people fleeing from the Southern states. However, her involvement in the war was not limited to medical work; Tubman also worked as a spy for the Union Army.

In 1863, she led the Combahee River Expedition in South Carolina, which freed more than 700 enslaved people. She was the first woman to lead a military operation in the United States, a remarkable feat that demonstrates her boldness and skill as a strategist. Throughout the war, she continued to gather key information and participate in risky missions, reaffirming her role as one of the most effective leaders of the abolitionist struggle.

After the war, Harriet Tubman grew up in Auburn, New York, where she continued her humanitarian work. She dedicated her life to social causes and worked tirelessly to support the rights of African Americans and women. She was an activist for women's rights to vote and collaborated with leading figures in the suffrage movement, such as Susan B. Anthony. In her later years, she developed a home for elderly and needy African Americans in Auburn, providing aid and support to those who needed it most.

Harriet Tubman's legacy remains a powerful inspiration for the civil and human rights movements. Her life represents the tireless fight for freedom and justice, and her courage has resonated in the struggles for racial equality and women's rights in the United States. Her name is synonymous with strength, resilience, and sacrifice in pursuit of freedom and human rights.

Today, her figure is widely recognized and honored. In 2016, the United States Department of the Treasury announced that her image would appear on the $20 bill, a significant tribute to her historic contribution to freedom and justice in the country. Harriet Tubman left an indelible legacy that continues to remind us of the importance of fighting for human rights and dignity for each one of us.

8. Marie Curie

Born Maria Salomea Skłodowska on November 7, 1867, in Warsaw, Poland, she was a pioneering scientist in the fields of physics and chemistry, known for her research into radioactivity. Her work transformed modern science and laid the groundwork for major advances in medicine, nuclear physics, and radiotherapy. She was the first woman to receive a Nobel Prize and the only person to win Nobels in two separate scientific fields, achieving one in physics (1903) and another in chemistry (1911). Her dedication to scientific knowledge and her contribution to humanity have left a monumental legacy.

Marie Curie excelled in the field of physics and chemistry, especially standing out in the study of radioactivity, a term she herself coined. Her interest in science began at an early age, influenced by her father, a science teacher in Poland. However, due to educational restrictions for women in her native country, she emigrated to Paris in 1891, where she entered the Sorbonne University to study physics and mathematics.

In 1894, she met physicist Pierre Curie, whom she married a year later and formed a legendary scientific collaboration. Together they began studying radioactive phenomena recently discovered by Henri Becquerel. In 1898, Marie and Pierre discovered two new elements: polonium, which Marie named after her native country, and radium. This discovery not only represented a fundamental advance in science but opened the doors to new areas of study in physics and medicine.

Marie Curie's contribution in this field earned her the Nobel Prize in Physics in 1903, which she shared with Pierre Curie and Henri Becquerel. After Pierre's death in 1906, Marie continued his work, demonstrating an unwavering dedication to scientific research. In 1911, she received her second Nobel Prize, this time in chemistry, in recognition of her discovery of radium and polonium and her isolation of pure radioactive substances – an impressive feat at the time.

Marie Curie's work in the field of radioactivity also had a profound impact on medicine. During World War I, mobile X-ray units, known as "petites Curies", were developed and used to treat wounded soldiers at the front. She worked in the field herself and trained other women to operate these units, providing them with an invaluable diagnostic service that helped save many lives.

In addition, her research was instrumental in the development of radiotherapy, one of the main therapies in the treatment of cancer. Her work enabled the application of radioactive elements to destroy cancer cells, making radium a revolutionary medical tool. Today, radiotherapy remains an important technique in oncology, and its use began largely thanks to Curie's pioneering work.

Marie Curie's legacy transcends the realm of scientific research; her life and achievements challenged social and gender barriers and paved the way for future generations of women in science. She was the first woman to hold a professorship at the Sorbonne University and the first to receive an honorary burial in the Pantheon in Paris, alongside great figures in French history. Her absolute dedication to science and her courage in challenging societal norms inspired countless women to pursue careers in scientific fields.

Marie Curie also founded the Curie Institute in Paris, a research center that today remains a benchmark in the study of physics and medicine. The institute has continued her work in cancer treatment and research, becoming a fundamental part of her scientific and humanitarian legacy.

Marie Curie remains a symbol of tenacity, dedication and brilliance. Her contribution to science not only changed our understanding of the world, but also paved the way for the use of science for the benefit of humanity. Through her discoveries and commitment to research, Marie Curie left a legacy that continues to influence modern medicine and scientific research.

Today, Marie Curie is remembered as one of the most brilliant scientific minds and a pioneer who broke gender barriers at a time when women had very limited access to science. Her life and work continue to be a model of excellence, sacrifice and commitment to the advancement of knowledge, leaving an indelible mark on the history of humanity.

9. Emmeline Pankhurst

Born on July 15, 1858, in Manchester, England, she was an undisputed leader of the British suffrage movement and one of the main figures in the fight for women's rights in the 20th century. Founder of the Women's Social and Political Union (WSPU), Pankhurst revolutionized the cause of women's suffrage with her direct and militant approach. Her determination and leadership made her a key figure in securing the right to vote for women in the United Kingdom, leaving a legacy in the history of civil and gender rights.

Emmeline Pankhurst excelled in the field of civil rights activism, focusing her life on women's suffrage. From an early age, Emmeline was influenced by her parents, who defended progressive ideals and exposed her to ideas about equality and justice. She married at the age of 20 to the lawyer Richard Pankhurst, who also supported women's suffrage and equal rights, which cemented her dedication to the cause of women's rights.

In 1903, after seeing the traditional suffragette movement making very slow progress, Pankhurst founded the WSPU with her daughters Christabel and Sylvia. The WSPU was notable for its motto "Deeds not words" and its radical approach. Pankhurst and her followers were determined to win the vote for women and used protest tactics that included marches, demonstrations, hunger strikes, and in some cases acts of civil disobedience, such as breaking windows and chaining themselves to public buildings. These tactics attracted public and media attention, as well as a harsh response from the government, which imprisoned numerous suffragettes, including Emmeline herself.

During their periods of imprisonment, Pankhurst and other suffragettes adopted hunger strikes as a form of protest. Authorities, seeking to avoid the deaths of prisoners and bad publicity, began force-feeding them, a painful and dangerous practice that further attracted public sympathy to the cause. In 1913, the British government enacted the Temporary Discharge of Prisoners Act (known as the "Cat and Mouse Act"), which allowed suffragettes on hunger strike to be released until they recovered their health, and then re-imprisoned. Despite these harsh measures, Pankhurst never gave up the fight and continued to organize campaigns and demonstrations.

With the outbreak of World War I in 1914, Emmeline Pankhurst suspended the activities of the WSPU and decided to support the war effort. She encouraged women to join the workforce and take on roles that had normally been filled by men. This change in circumstances caused British society to begin to value the contributions of women, and the British government reconsidered its policies towards women's suffrage.

Eventually, in 1918, Parliament passed the Representation of the People Act, which granted voting rights to women over the age of 30 who met certain property requirements, and in 1928, suffrage was extended to all women over the age of 21 on equal terms with men. Although Emmeline Pankhurst did not live to see full suffrage, which was passed shortly after her death, her tireless struggle was instrumental in achieving this goal.

Emmeline Pankhurst's legacy can be seen today in the advancements of civil and gender rights, inspiring generations of women to fight for equality and democratic rights. She was a revolutionary figure who proved that persistence and resistance could transform the social and political landscape of her country and the world. Her militant approach and leadership skills make her one of the most influential figures in the history of feminism and women's rights.

In 1928, shortly before her death, Emmeline was appointed as a parliamentary candidate for the Conservative Party, which demonstrates the recognition she achieved at the end of her life. In 1930, two years after her death, a statue was erected in her honour near the Palace of Westminster in London, and her legacy continues to be an inspiration to those fighting for justice and gender equality.

Emmeline Pankhurst not only changed the history of the United Kingdom, but also left a lasting impact on the women's rights movement globally. Her life is remembered as a testament to courage and perseverance in the pursuit of justice and equality, and her legacy lives on in current struggles for women's rights around the world.

10. Frida Kahlo

Born on July 6, 1907, in Coyoacán, Mexico City, she was a uniquely talented painter and one of the most recognized and admired artists of the 20th century. Her work, marked by her personal life and deep symbolism, gave her a prominent place in modern art and the surrealist movement. Kahlo used her art to express her physical and emotional pains, becoming an icon of authenticity, female empowerment, and resistance. Although she lived a life full of physical suffering, her creativity and authenticity led her to leave a legacy in art and culture.

At the age of six, Kahlo contracted polio, which permanently affected her right leg. However, it was at the age of 18 that her life changed radically due to a bus accident that caused multiple fractures in her spine, pelvis, and other parts of her body. The severe injuries forced her to spend long periods in bed, where she began to paint to escape her suffering. Her mother installed a mirror over her bed, allowing her to use her own image as a model. This marked the beginning of her distinctive style, using her own figure to express her feelings and experiences.

In 1929, Kahlo married the renowned Mexican muralist Diego Rivera, a stormy and passionate relationship that deeply influenced her work. Both shared political ideals, identifying with communism and Mexican nationalism, which was also reflected in the iconography of her paintings.

Frida Kahlo excelled in the field of painting, creating a unique style that combined surrealism with realism. Although André Breton, one of the founders of surrealism, considered Kahlo to be a surrealist, she herself denied belonging to this movement, stating that she did not paint dreams, but her reality. Her work is characterized by deep symbolism and heartbreaking sincerity, reflecting her internal struggles, her complex relationship with Rivera, her love for Mexico, and her physical and emotional pain.

Many of her self-portraits depict themes such as duality, identity, suffering and femininity, rendered in vibrant colours, indigenous Mexican symbols and references to folklore. Paintings such as The Two Fridas (1939) and The Broken Column (1944) are emblematic of her style, showing Kahlo in all her vulnerability and strength, facing her pain and emotional losses. Her works not only explore her personal life, but also issues of gender, identity and internal struggle, becoming a visual language to express the complexity of being a woman and living in a body that she herself describes as a "cage".

Kahlo also stood out as a political activist, advocating for communism and women's rights at a time when female voices were disregarded in Mexican society and in the art world. Alongside Diego Rivera, Frida was committed to social causes, supporting the fight for equality and justice. Her way of dressing and her personal life, including her traditional Tehuana attire and her open expression of her sexuality, challenged conventional norms of gender and femininity, and remains a source of inspiration for contemporary feminism.

Frida Kahlo's legacy is immense and multifaceted, as she not only revolutionized modern painting, but also became a cultural icon and a symbol of resistance. Her home in Coyoacán, the famous "Blue House," was converted into a museum in 1958, four years after her death, and continues to be one of the most visited museums in Mexico, attracting admirers from all over the world.

Through her art, Kahlo helped open spaces for dialogue on issues such as identity, the female body, and pain, themes that are still explored by artists and activists today. In addition, her image is recognized globally, becoming a reference figure for movements of cultural identity, feminism, and resilience.

Kahlo's image has been reinterpreted in multiple disciplines, from fashion to music to film, and her life has been the subject of numerous films, documentaries and books. In 2002, her life was brought to the big screen in the movie Frida, starring Salma Hayek, which further contributed to her international recognition. Kahlo remains an inspiration to those seeking authenticity and courage to express their own pain and inner beauty.

Frida Kahlo did not only excel as an artist; her life and work continue to be a symbol of the human capacity to transform pain into a creative force. Her legacy transcends art and represents a personal and collective struggle for self-acceptance, identity and empowerment, leaving an indelible mark on culture and art history.

11. Rosalind Franklin

Born on July 25, 1920, in London, England, she was a prominent British chemist and crystallographer whose research was crucial to the discovery of the structure of DNA. Despite the barriers she faced in her time as a woman in the scientific world, her dedication and precision in using X-ray diffraction allowed her to capture detailed images of DNA, making essential contributions to the understanding of molecular biology. Her work on DNA, the coal bug, and the tobacco mosaic virus cemented her reputation as a scientific pioneer.

From a young age, Franklin demonstrated outstanding talent for science and unwavering determination. She was educated at prestigious institutions and earned her PhD in physical chemistry from the University of Cambridge in 1945. During and after World War II, she worked in Paris, where she honed her skills in using the X-ray diffraction technique at the Laboratoire Central des Services Chimiques de l'État. This technique would become the main tool of his most important investigations, allowing him to visualize molecular structures with great precision.

In 1951, Franklin returned to England and began working at King's College London. Here, in John Randall's laboratory, he focused on DNA research. Franklin and his student Raymond Gosling managed to produce the first clear images of DNA in 1952, known as the 51 photographs, which revealed a helical structure in the DNA molecule. This image would become the fundamental basis for the later discovery of the double helix structure of DNA.

The field in which Franklin excelled was X-ray crystallography, and his work on DNA was transcendental. Through his research, he discovered that DNA had a helical structure, with a repetitive pattern that suggested the presence of a double helix. However, her work was overshadowed by a controversy: her colleague Maurice Wilkins, without her permission, showed the famous photograph 51 to James Watson and Francis Crick, who used this key information to develop their double helix model. In 1953, Watson and Crick published their model in the journal Nature, indirectly mentioning Franklin's work, which did not receive the same recognition in her time.

In 1962, Watson, Crick and Wilkins were awarded the Nobel Prize in Physiology or Medicine for the discovery of the structure of DNA. Franklin, who had died in 1958 at the age of 37 from ovarian cancer, was not considered for the prize, in part because the Nobel Prize is not awarded posthumously. Despite this, her contribution is today widely recognized and valued in the scientific community.

Following her time at King's College, Franklin continued her research in J.D. Bernal's laboratory at Birkbeck College, where she studied viruses, including tobacco mosaic viruses. Her work in virology was equally groundbreaking, laying the groundwork for virus crystallography and contributing to the development of molecular biology. She published significant papers on the structure of viruses, which were fundamental to the study of viral diseases in humans and plants.

Her work helped shape the field of structural biology, and, although short-lived, her research in crystallography and virology continued to impact science in profound ways.

Rosalind Franklin's legacy is immense and has only grown stronger over time. Today, she is recognized as a pioneer in molecular biology and a symbol of gender equality in science. Her contributions to X-ray crystallography and her rigorous scientific methodology have been essential to the modern understanding of genetics and molecular biology.

Institutions around the world have honored her memory by naming buildings, scholarships, and laboratories after her, and NASA named the Mars rover on its ExoMars 2022 mission Rosalind Franklin in recognition of her contributions. Today, her story serves as an inspiration to women who wish to pursue science and a reminder of the crucial role they will play in fields where the female presence has historically been limited.

12. Simone de Beauvoir

Born on January 9, 1908, in Paris, France, she was a prominent philosopher, writer, and activist who left a fundamental impact on 20th-century philosophy, literature, and feminism. Known for her most influential work, The Second Sex (1949), de Beauvoir analyzed the roots of women's oppression and offered a profound critique of patriarchal society, positioning herself as one of the precursors of modern feminism. Her life and work were not only limited to theoretical concepts, but she also defended the freedom and autonomy of women in a world dominated by gender restrictions and traditional norms.

From a young age, de Beauvoir demonstrated exceptional intelligence and a strong inclination towards the study of philosophy and literature. She graduated in philosophy from the Sorbonne University and then entered the École Normale Supérieure, one of the most prestigious academic institutions in France. It was there that she met Jean-Paul Sartre, with whom she would establish a unique intimate and professional relationship. Although they remained in an open relationship, they shared existentialist ideals, and a vision committed to personal and social freedom.

De Beauvoir stood out in the field of existentialism, a philosophical movement that defended individual freedom and personal responsibility. Her relationship with Sartre was more than a romantic connection; it was an intellectual collaboration and an opportunity to develop revolutionary ideas about autonomy and freedom, fundamental themes in her work. Both considered that existence precedes essence, which means that the human being defines himself through his actions and decisions. This notion significantly influenced the way in which de Beauvoir approached female oppression, arguing that women could build their identity and find freedom through their choices.

The area in which Simone de Beauvoir stood out the most was in feminist theory. Her work The Second Sex, published in two volumes in 1949, is a comprehensive analysis of women's oppression and one of the first systematic explorations of the causes and effects of gender inequality. In this work, de Beauvoir described how women had historically been relegated to a secondary status, categorized as "the other" in relation to men. The famous phrase "One is not born a woman, one becomes one" sums up her theory that femininity is a social construct and not an immutable biological condition.

In The Second Sex, de Beauvoir addressed issues such as marriage, motherhood, work, sexuality and individual freedom. She questioned the idea that women's roles were limited to the home and family and urged women to challenge traditional norms that oppressed them. This work marked the beginning of a radical critique of patriarchy and became a pillar of modern feminism, influencing women's liberation movements around the world.

Beyond her theoretical contribution, de Beauvoir was also a committed activist for the rights of women and oppressed groups. In the 1970s, she participated in protests and the fight for the legalization of abortion in France, winning support and approval of the voluntary termination of pregnancy law in 1975. She was also an advocate for women's labor rights, equal access to education, and equal pay.

Simone de Beauvoir's legacy is broad and multifaceted. She was a pioneer in the construction of contemporary feminism, not only promoting equality, but also redefining female identity from a perspective of freedom and autonomy. Her work has deeply influenced feminist theories and political activism and remains essential reading for those studying gender, ethics, and philosophy.

Beauvoir's influence is especially palpable in current discussions about gender, sexuality, and human rights. The Second Sex has been translated into multiple languages and is considered one of the most important and cited texts in gender studies. It inspired and continues to inspire millions of women and men to question gender norms and recognize the impact of patriarchy on their lives.

13. Margaret Thatcher

Born on October 13, 1925, in Grantham, England, she was a British politician notable for her role as Prime Minister of the United Kingdom from 1979 to 1990. As leader of the Conservative Party and the first woman to hold this office in her country, Thatcher left a profound mark on British and world politics. Known as the "Iron Lady" for her firm decision-making, Thatcher pushed through economic and social reforms that profoundly transformed the United Kingdom and defined an era of modern conservatism.

From an early age, Thatcher showed an interest in politics and economics. The daughter of a local shopkeeper and politician, she learned from her father the value of self-reliance and personal effort. She studied chemistry at Oxford University, where she was president of the university's Conservative Association. After graduating, she devoted herself to the study of law and briefly worked as a lawyer.

Her entry into politics occurred in 1959, when she was elected as a member of Parliament for the Conservative Party. During the 1960s and 1970s, Thatcher rose through the party ranks to become Minister of Education under the Conservative government of Edward Heath. In 1975, after the Conservatives' failure in the elections, Thatcher was elected leader of the Conservative Party, positioning herself as the first woman to lead a major party in the United Kingdom.

Thatcher excelled in the field of economic policies and the transformation of the British economic model, developing what is known as "Thatcherism". In 1979, she became the first woman to assume the office of Prime Minister, at a time when the United Kingdom was going through a serious economic crisis, with high inflation, unemployment and a strong union power that generated frequent labor conflicts.

Thatcherism promoted a neoliberal approach characterized by reduced state intervention in the economy, privatization of public enterprises, deregulation of financial markets, and reduction of the power of unions. Thatcher believed in a "small state" that allowed free competition and individual responsibility. Her administration implemented cuts in social spending and promoted a free market-based economy. These policies faced much resistance, especially from the working class and unions, but they succeeded in reducing inflation and returning the country to international economic competitiveness.

Thatcher also implemented the privatization of state-owned companies, such as British Telecom and British Airways, and reformed the tax system. Although her economic reforms were controversial and caused unemployment in several industries, especially in the industrial regions of northern England, they also contributed to revitalizing the British economy in the 1980s, generating an economic boom and transforming London into a global financial center.

In addition to her impact on the economy, Thatcher excelled in the field of foreign policy. One of the most significant moments of her leadership was the

Falklands War in 1982, a military conflict with Argentina over control of the Falkland Islands. Thatcher's swift decision to send British forces to recapture the islands after the Argentine invasion was supported by most of the British people, strengthening her image as a determined and patriotic leader.

Thatcher's relationship with the United States was also very close, especially with President Ronald Reagan. Both shared a conservative worldview and supported similar policies. Thatcher was one of the strongest voices against the expansion of Soviet influence and strongly supported Reagan in the Cold War, defending the alliance with NATO and promoting a message of resistance to the Soviet Union.

Margaret Thatcher's legacy is complex and, for many, polarizing. In the economic realm, her free market and privatization policies transformed Britain into a more market-oriented economy and marked the end of the era of social democratic consensus that had characterized the post-war period. While they succeeded in strengthening the British economy, her policies also generated high levels of inequality and marginalized certain sections of the population. The lack of support for the working class and industrial communities left a mark of resentment, especially in the north of England and Scotland.

Thatcher's figure remains a source of controversy in British politics. For her supporters, she was a visionary leader who developed economic vitality in the United Kingdom and successfully challenged union and state influence. For her detractors, her policies had a very high social cost, weakening social cohesion and increasing inequality.

In 1990, Thatcher resigned from office after losing the support of her party, in part due to her advocacy of the controversial "poll tax" and internal divisions over the relationship with the European Community. Still, her influence on the Conservative Party and British politics was long-lasting, and her ideas were adopted by numerous later leaders.

Thatcher died on April 8, 2013, at the age of 87, leaving a lasting impact that continues to generate debate about the role of the state, the free market, and social justice.

14. Malala Yousafzai

Born on July 12, 1997, in the Swat Valley, Pakistan, she is an internationally recognized activist for her advocacy of girls' education and human rights. Her courage in the face of extreme threats in her fight for education has made her a global inspiration and a symbol of resistance to oppression. At just 17 years old, Malala was awarded the Nobel Peace Prize, becoming the youngest person to receive this prestigious award. Her story is a reminder of the power of youth and the impact activism can have on social transformation.

Malala grew up in the Swat Valley, a mountainous region in northern Pakistan, in a family that deeply valued education. Her father, Ziauddin Yousafzai, was an educator and activist who ran a girls' school in his hometown and encouraged Malala to express herself and learn without limitations – something uncommon in a region where women's rights were highly restricted.

However, the social and political context in which Malala lived was difficult and dangerous. Beginning in 2007, the extremist group Taliban began to gain control in the Swat Valley, imposing strict restrictions and banning female education. Taliban influence in the region meant the closure of several girls' schools and increased threats against those trying to advocate for female education.

From a young age, Malala was committed to the cause of education and spoke out against the ban imposed by the Taliban. At the age of 11, she began writing a blog in Urdu under a pseudonym for the BBC, chronicling her life in Swat under Taliban occupation. In her writings, Malala described the difficulties in accessing education and denounced the injustice of depriving girls of their right to learn.

On 9 October 2012, as Malala was returning home on a school bus, she was attacked by a Taliban member who shot her, seriously wounding her in the head and neck. Her life was in great danger, and the attack sparked a wave of global outrage. Malala was flown to the UK, where she received medical treatment and eventually recovered.

Despite the threats and the attack, Malala was undeterred. Instead of walking away from activism, she redoubled her commitment to female education and the defence of human rights. In 2013, she published her autobiography, I Am Malala, a book that recounted her story and her fight for access to education. That same year, she co-founded the Malala Fund with her father, an organization dedicated to promoting education for girls and working to improve access to education in communities around the world.

In 2014, at the age of 17, Malala was awarded the Nobel Peace Prize, becoming the youngest person to receive this recognition. The Nobel Committee's decision highlighted her courage and the power of her activism. She shared the prize with Kailash Satyarthi, an Indian activist against child labor, sending a message of unity between two countries, India and Pakistan, regularly at war.

Since receiving the Nobel Prize, Malala has continued her studies and her work in education advocacy. In 2017, she entered the University of Oxford, where she studied philosophy, politics and economics, a background that complements her experience as an activist. Through the Malala Fund, she has promoted projects in countries such as Syria, Nigeria, Brazil and Afghanistan, where access to education is hampered by armed conflict, economic crisis and gender barriers.

In addition to her work with the Malala Fund, she has spoken at international forums and collaborated with world leaders to promote access to education as a fundamental human right. In 2015, she was invited to address the United Nations and, in her speech, she highlighted the importance of investing in education to create a more just and peaceful world.

Malala Yousafzai's legacy focuses on the field of human rights and, particularly, on the fight for girls' education. Her story is an example of resilience and courage, and her influence has inspired millions of people around the world to support gender equality and defend educational rights. Thanks to her efforts, access to education has become an international priority, and her figure has helped to raise awareness of the problems faced by girls in situations of conflict and extreme poverty.

Malala continues to work so that every girl can receive an education and live without fear, and her story continues to inspire new generations to fight for a more equitable and fairer world.

15. Angela Merkel

Born on July 17, 1954, in Hamburg, Germany, she is a prominent German politician and European leader who specialized as Chancellor of Germany from 2005 to 2021. Her leadership was fundamental in times of great challenge for Europe and the world, and she is recognized for her pragmatic approach, her stability in decision-making and her commitment to international cooperation. Merkel was the first woman to serve as chancellor in Germany and left a legacy that established her country as an economic and political power in Europe.

Before entering politics, Merkel had a distinguished career in science. She was born in western Germany but grew up in East Germany (the German Democratic Republic) after her family moved to the communist zone when she was a child. She studied physics at the University of Leipzig and earned a doctorate in quantum chemistry in 1986. During her years of study and work in scientific research, Merkel was known for her intelligence and academic rigor, which laid a solid foundation for her analytical style in politics.

Merkel's political career began shortly before the fall of the Berlin Wall in 1989, an event that marked a decisive moment in her life and in the history of Germany. She entered politics during the process of German reunification, joining the Democratic Alliance party in East Germany, which was later integrated with the Christian Democratic Union (CDU). In 1990, she was elected to the Bundestag (German Parliament) and quickly rose in the party under the leadership of Helmut Kohl, who appointed her Minister for Women and Youth, and later Minister for the Environment.

Her position in the CDU strengthened over the years, and in 2000 she was elected party leader. This was a significant victory in a traditionally male-dominated party and allowed Merkel to consolidate her political leadership. Finally, in 2005, she became Chancellor of Germany, beginning a 16-year career at the head of the German government, making her one of the most influential European leaders of the 21st century.

During her tenure as chancellor, Merkel excelled in the field of international politics and crisis management. Her pragmatic approach, focused on stability and consensus, helped her face several economic and political challenges that marked her leadership.

One of her first challenges was the 2008 global financial crisis, in which Merkel played a pivotal role in designing economic policies to protect Germany from the worst consequences of the recession. She was also a key figure in managing the debt crisis in the Eurozone, advocating a policy of austerity in EU countries facing economic difficulties, such as Greece, Spain and Italy. Although these policies were controversial, Merkel maintained that they were necessary to ensure the stability of the Eurozone and protect the euro.

Merkel also led Germany's and the European Union's response during the 2015 migration crisis, when she decided to open Germany's doors to refugees from Syria and other conflict zones. This decision was an unprecedented humanitarian act, but it also created divisions at home and in Europe. Her stance on migration showed her commitment to human rights and her values of solidarity, but it also led to a rise in support for nationalist and anti-immigration parties in Germany.

In addition to her influence in Europe, Merkel played a crucial role in global diplomacy. She maintained a close relationship with the United States, although with a critical stance, when necessary, especially during the administration of Donald Trump. An advocate of multilateral cooperation, Merkel was one of the main voices in support of climate change, pushing environmental policies both at the national and European level.

Merkel also worked closely with international leaders such as Vladimir Putin and Xi Jinping, maintaining a balance between defending democratic values and the need for economic and diplomatic cooperation. Her approach led her to earn the respect of the international community and consolidate Germany as a leading country in Europe and an influential voice on global issues.

Angela Merkel's legacy is characterized by the stability and solidity she brought to Germany during 16 years of mandate. Her leadership in a period of constant change and challenges, from financial crises to geopolitical conflicts, strengthened Germany's position as Europe's leader. Merkel is seen as a model of pragmatic and rational leadership, and her reserved style, often described as modest and cautious, was a stark contrast to the style of other, more polarizing leaders.

Under her leadership, Germany established itself as Europe's largest economy and a pivotal political force in the European Union. Merkel left office in 2021, after four consecutive terms, having guided her country through critical moments and upheld the values of European cooperation and stability.

Affectionately known as "Mutti" (mother) to her supporters, Merkel represented a figure of stability and confidence. Although she rarely showed emotion in public, her empathetic approach to citizens' problems, especially during the migration crisis, projected an image of an approachable and approachable leader. Her discreet style and ability to listen were qualities that marked her career and her way of governing.

16. Hypatia of Alexandria

She was a philosopher, mathematician and astronomer from ancient Alexandria, Egypt, renowned for her profound knowledge and her ability to teach and share ideas at a time when women had very limited access to knowledge. Hypatia excelled in the field of science and philosophy, contributing to the development of mathematics and astronomy and becoming one of the most prominent intellectual figures of her time. Her life and tragic end have made her a symbol of the love of knowledge and freedom of thought, leaving a legacy in the history of science and rational thought.

Hypatia was born around 370 AD in Alexandria, one of the most vibrant and important cities of the ancient world, known for its famous Library and for being a center of knowledge and culture. Her father, Theon of Alexandria, was a renowned mathematician and astronomer who taught Hypatia from an early age, encouraging her curiosity and preparing her in disciplines such as mathematics, astronomy, and philosophy. Under her father's tutelage, Hypatia studied not only the teachings of the great Greek mathematicians, but also the ideas of philosophers such as Plato and Aristotle.

Her education in Alexandria gave her access to knowledge of civilizations such as the Greek, Egyptian, and Roman, which contributed to her unique perspective and deep understanding of different schools of thought. As she progressed in her studies, Hypatia became noted for her skill and precision in mathematics and philosophy, earning the respect of her contemporaries.

Hypatia became an influential teacher and scholar at the Neoplatonic school in Alexandria, where she taught mathematics, astronomy, and philosophy. Her teaching approach was widely respected, and her classes attracted students from various parts of the ancient world, many of whom were of Christian background, although she herself practiced paganism. Hypatia taught and wrote on subjects such as geometry, algebra, and astronomy, and is believed to have developed and improved instruments such as the astrolabe and hydrometer, used in observing the stars and measuring liquids.

In addition to her contributions to the exact sciences, Hypatia was noted for her critical thinking and rational approach, promoting a philosophical approach to life based on logic and observation. She was a central figure in the Neoplatonic movement, which sought to reconcile Plato's ideas with spirituality and science, promoting a search for truth through knowledge and reason.

Although none of her works have survived directly, Hypatia is known to have written commentaries on the works of mathematicians such as Diophantus of Alexandria and Apollonius of Perga, indicating her deep knowledge of geometry and algebra. Her commentaries on Diophantus contributed to the understanding of number theory and equations. She is also thought to have written on Euclid's Elements, helping to preserve and clarify the mathematical teachings of the Greek world.

In astronomy, she is said to have worked on planetary models and calculations related to the motion of the stars. Although the extent of her original contributions is not known for certain, her teaching and preservation of this knowledge helped keep these ideas alive at a time when science was beginning to be influenced and limited by religious doctrines.

During Hypatia's lifetime, Alexandria was a city in tension due to the conflict between the followers of the rapidly expanding Christian faith and those who still followed pagan philosophy. Hypatia, being a Neoplatonist and an advocate of secular learning, found herself in a vulnerable position. She was close to Orestes, the Roman prefect of Alexandria, who represented a stance of tolerance towards paganism and sought to limit the power of Bishop Cyril, leader of the city's growing Christian community.

This conflict culminated in 415 AD, when a mob of radical Christians, encouraged by political and religious tensions, attacked Hypatia. She was brutally murdered on the streets of Alexandria, and her death has been interpreted as an act of violence against rational thought and secular knowledge. Her assassination shocked the intellectual community of her time and symbolically marked the decline of the ancient scientific and philosophical tradition in Alexandria.

Despite her tragic end, Hypatia left a legacy in the history of science and philosophy. She is remembered not only for her intellectual achievements, but also for her courage in staying true to her ideals and dedicating herself to the pursuit of truth and knowledge in a time of great hostility. Over time, Hypatia has become a symbol of freedom of thought and resistance in the face of fanaticism. Her life has inspired generations of thinkers and scientists, and her story remains a reminder of the importance of reason and science in society.

17. Hatshepsut

She was one of the most prominent female pharaohs in the history of Egypt, ruling during the New Kingdom period (18th Dynasty). She was the first woman to be proclaimed pharaoh in her own right and not just as regent, something extraordinary in a system traditionally dominated by men. Hatshepsut excelled in the expansion of the economy and the peace policies she implemented, as well as carrying out large-scale architectural projects that symbolized the prosperity of her government and her consolidation as one of the most prosperous and powerful rulers of ancient Egypt.

Hatshepsut was born in Thebes around 1507 BC, daughter of Pharaoh Tuthmosis I and his main wife, Ahmose. Upon the death of her father, her half-brother and husband, Thutmose II, ascended the throne and Hatshepsut became his Great Royal Wife, but following his early death, Egypt came under Hatshepsut's regency on behalf of the young Thutmose III, the designated successor, who was still a child.

Initially, Hatshepsut ruled as regent, but eventually took the title of pharaoh, something unprecedented for a woman in Egyptian history. Determined to establish herself as sovereign, Hatshepsut adopted the symbolism and rituals of the pharaohs, even depicting herself with the ceremonial beard and pharaoh's headdress in official inscriptions, and demanding to be called "Son of Ra." This not only legitimized her position, but also underlined her intention to rule with the same authority as any male pharaoh.

During the roughly 22 years of her reign, Hatshepsut focused her efforts on strengthening the Egyptian economy and expanding trade. Rather than focusing on war and conquest, as many of her predecessors had done, Hatshepsut promoted peace and stability within Egypt's borders, resulting in a period of prosperity.

One of her greatest achievements was the trading expedition to the legendary kingdom of Punt (probably in the modern-day Horn of Africa region). This mission brought back to Egypt a wealth of riches, including frankincense, myrrh, ebony, ivory, and exotic animals. Inscriptions and reliefs in her famous mortuary temple at Deir el-Bahari detail this journey, highlighting the success of her trade policies and the wealth they brought to Egypt. Hatshepsut used these resources to beautify the country, investing in the development of temples, monuments, and statues, marking her era as one of abundance and splendor.

Hatshepsut left behind an impressive architectural legacy, most notably the mortuary temple at Deir el-Bahari, considered a masterpiece of ancient Egyptian architecture. Located on the west bank of the Nile, opposite Thebes, this temple is a monumental structure of stepped terraces harmoniously integrated with the surrounding natural landscape. Decorated with reliefs and statues of Hatshepsut, the temple narrates aspects of her reign, her divine lineage, and the famous expedition to Punt.

This architectural project not only reflected her ambition and ability to direct large construction projects, but also cemented her place in history as a visionary builder. Her architectural achievements at Deir el-Bahari and other constructions at Karnak and other Egyptian cities demonstrate her dedication to the cultural and religious development of her nation, elevating Egyptian architecture to new levels of sophistication and symbolism.

At the end of his reign, Thutmose III fully assumed the throne and embarked on a campaign to remove traces of Hatshepsut from the historical record. To consolidate his own legitimacy, Thutmose III is believed to have ordered many of her statues and monuments destroyed or erased, and her image and name removed from royal lists. However, although some of her monuments were altered and her inscriptions changed, Hatshepsut's legacy as one of Egypt's most innovative and successful rulers survived.

Hatshepsut excelled in politics and architecture, and her reign marks a period of economic and cultural splendor. Her ability to establish a strong government, the prosperity Egypt achieved under her rule, and her architectural achievements contributed to her being recognized as one of history's great pharaohs.

Hatshepsut left a legacy that challenged the gender stereotypes of her time and paved a path for future female rulers in world history. She is remembered today not only as one of the few women to rule as a pharaoh in ancient Egypt, but also as an example of determination, intelligence, and leadership.

18. Florence Nightingale

She was a British nurse, social reformer, and pioneer of modern nursing, whose work laid the groundwork for principles and practices still used today. Known as "The Lady with the Lamp" due to her dedication and efforts during the Crimean War, Nightingale excelled in the field of medical care and hospital hygiene, revolutionizing nursing and managing to improve conditions in hospitals. Her legacy extends to the creation of care protocols that saved countless lives and her influence in the creation of nursing as a dignified and professional profession.

Florence Nightingale was born on May 12, 1820, in Florence, Italy, to a wealthy English family. From a young age, Florence displayed great intelligence and was well-versed in mathematics, science, and literature, which was unusual for women of her time. Despite her family's expectations that she led a social life as a society lady, Florence developed a deep calling to serve others, which she described as a "calling from God." At the age of 24, she decided to become a nurse, an unusual and controversial choice, as at the time nursing was a low-status job reserved for lower-class women.

In 1853, the Crimean War broke out, and the plight of wounded soldiers, cared for in appalling conditions, reached the ears of the public in Britain. Mortality rates in field hospitals reached 40%, and the lack of hygiene and infrastructure meant that many soldiers died from infections rather than from combat wounds. Faced with this crisis, Nightingale was asked to lead a group of nurses to improve care on the war front.

Florence and her team arrived at the hospital in Scutari in 1854, where they were confronted with an unsanitary and chaotic environment. She immediately implemented radical changes, including clean facilities, access to clean water, handwashing, and proper ventilation. Through these basic hygiene measures, she managed to reduce the mortality rate at the field hospital from 40% to just 2%. Her dedication to wounded soldiers, walking the wards at night with a lamp, earned her the affectionate nickname “The Lady with the Lamp.”

In addition to her direct work with patients, Nightingale pioneered the use of statistics to demonstrate the importance of hygiene in medical care. She created innovative charts (such as the polar area chart, also called the “rose diagram”), which visually presented mortality data before and after her reforms. These charts helped to make it clear that unsanitary conditions were the leading cause of hospital deaths, and her findings were instrumental in convincing authorities and the public of the need for reforms in the hospital system.

Upon her return to Britain, Nightingale continued to promote improvements in public health and military hospitals. In 1859, she published Notes on Nursing, a work that set out the basic principles of patient care and became an essential reference for future generations of nurses.

In 1860, with funding granted by the British government in recognition of her work, Florence founded the Nightingale Training School for Nurses at St. Thomas' Hospital in London. This school was the first to formalize nursing education, establishing a structured curriculum and professional standards for nursing practice. Under Florence's vision, nursing was transformed from a precarious activity into a prestigious profession, with a focus on empathy, technical competence, and hygiene.

The school was an immediate success and became a model for nursing programs around the world.

Graduates of the Nightingale School took the principles of modern nursing to other hospitals and countries, spreading their legacy globally. The nursing profession, once devalued, began to gain respect and dignity, becoming an essential part of health care systems.

Florence Nightingale wrote numerous books and articles on healthcare throughout her life, and her influence reached into political and social spheres. In 1883, Queen Victoria awarded her the Royal Red Cross in honour of her services, and in 1907, she became the first woman to receive the Order of Merit, one of the highest honours in the United Kingdom.

Her legacy lives on to this day: the professional model of nursing, hygiene practices in hospitals and the use of statistics in public health are all pillars that Florence Nightingale helped to establish. Her influence is celebrated each year on May 12, her birthday, on International Nurses Day, which honours the dedication and service of nurses around the world.

She is remembered as one of the most influential figures in modern medicine and a symbol of compassion, dedication and service.
19. Indira Gandhi

She was the first and only woman to be Prime Minister of India and is remembered as one of the most influential political figures in her country. She excelled in the field of politics and governance, achieving important advances in the modernization of India and facing complex challenges of unity and stability in a country with enormous cultural and social diversity. Her leadership in times of crisis and her decisions in areas such as the economy and national security defined her legacy as one of the most outstanding politicians of her time, although she was also a controversial figure.

Indira Priyadarshini Gandhi was born on November 19, 1917, in Allahabad, India, into a family of prominent political leaders. She was the daughter of Jawaharlal Nehru, the first minister of independent India, which exposed her from a young age to an environment of political activism and commitment to the Indian nationalist cause. Her education was of international standard, as she studied at Oxford University and returned to India with a strong determination to follow her father's path and contribute to the building of a modern, independent India.

Indira joined the Congress Party and, following the death of Lal Bahadur Shastri, assumed the office of Prime Minister in 1966. During her first term, she faced several important challenges, including hunger and food crisis, economic growth, and tensions with neighboring countries. One of her most important achievements was the implementation of the Green Revolution in India, an agricultural modernization program that significantly increased food production and reduced the country's dependence on imported staple grains. This program improved food security and raised the country's self-sufficiency, making a lasting impact on the rural economy.

In 1971, during her second term, Indira led India in the war against Pakistan, supporting the Bangladesh independence movement. Her decisive leadership and backing of Bangladesh's independence strengthened India's image as a regional power and cemented Gandhi's position as a strong, nationalist leader.

Although Indira Gandhi is renowned for her achievements, she was also a controversial political figure due to her implementation of the "State of Emergency" between 1975 and 1977. The declaration of emergency was a response to growing opposition and political instability, allowing Gandhi to rule with extraordinary powers and suspension of civil rights. During this period, her birth control policies, forced evictions, and media censorship drew criticism both in India and abroad. The move sparked a strong social backlash as it was seen as a restriction on democracy and led to her defeat in the 1977 elections.

However, in 1980, after a period of instability and discontent with the government of her successors, Indira Gandhi was re-elected as Prime Minister. During this second term, Gandhi faced separatist challenges in Punjab, where conflicts with radicalized Sikh groups led to a situation of high tension in the country.

One of the most critical moments of her mandate was Operation Blue Star in 1984, a military intervention in the Golden Temple in Amritsar, the most sacred site for the Sikh community, to dislodge armed separatists who had taken refuge there. The operation resulted in many deaths and injuries, as well as damage to the temple, and generated deep resentment among Sikhs. In retaliation, on October 31, 1984, Indira Gandhi was assassinated by two of her Sikh bodyguards at her residence in New Delhi.

Indira Gandhi left a complex, but deeply significant legacy in the history of India. She is remembered as a visionary leader who modernized the country, promoting food self-sufficiency policies, strengthening national defense and helping India emerge as a regional power. Her decision to lead the intervention in Bangladesh and her emphasis on nationalism helped India consolidate its position in South Asia.

On the other hand, her authoritarian policies and the period of the State of Emergency left a wound in India's democratic legacy, causing internal divisions and questions about her government. However, her ability to face difficult situations and make high-impact decisions won her the admiration of millions of people, and she remains a symbol of female leadership in a context historically dominated by men.

20. Rigoberta Menchú

She is a Guatemalan activist, defender of indigenous rights and winner of the Nobel Peace Prize in 1992. She excelled in the field of human rights, especially in the defense of the rights of indigenous peoples in Guatemala and throughout Latin America. Through her peaceful struggle and her voice in the international arena, she has made visible the injustices and abuses suffered by indigenous communities and has promoted equality and cultural respect. Her legacy stands out in the creation of a global platform for indigenous struggles and in the inspiration of future generations of activists and leaders.

Rigoberta Menchú was born on January 9, 1959, to an indigenous Quiché family in the village of Laj Chimel, in the department of Quiché, Guatemala. Her childhood was marked by the poverty and inequality that affected indigenous communities in her country, as well as by the discrimination and abuse they suffered in the context of a society deeply divided by ethnic and economic reasons. Rigoberta witnessed from a young age the struggle of her family and her community for land rights and for better living conditions, which instilled in her a strong commitment to justice.

In the 1970s, during the Guatemalan civil war, the situation of indigenous peoples worsened. Her family and community were victims of violence and repression, culminating in the death of her father, Vicente Menchú, and other members of her family at the hands of the Guatemalan army. These tragic events prompted Rigoberta to take an active and public position in defense of her people and against the atrocities experienced by indigenous peoples in Guatemala.

In 1982, Rigoberta Menchú achieved international notoriety with the publication of her autobiography, My Name is Rigoberta Menchú and That's How My Consciousness Was Born, written in collaboration with anthropologist Elisabeth Burgos-Debray. In this book, Rigoberta describes the harsh living conditions of indigenous Guatemalans, the violence suffered by her family and community, and the atrocities committed during the armed conflict in Guatemala. The book was instrumental in raising awareness of the injustices and abuses in Guatemala and in raising international public awareness of the causes of indigenous peoples and their rights.

The publication of her story was a powerful tool that exposed the problems of repression and discrimination in Latin America. The international visibility she gained fueled a movement of solidarity and support for indigenous communities in Guatemala, and Rigoberta became one of the leading voices in the fight for social justice.

In 1992, Rigoberta Menchú received the Nobel Peace Prize in recognition of her work for social justice and the rights of indigenous peoples. This distinction gave her a global platform to denounce repression and discrimination and positioned her as a symbol of indigenous resistance and dignity. With this recognition, Rigoberta managed not only to denounce injustices in her own country, but also to inspire indigenous communities and activists around the world to defend their rights and cultural identity.

After receiving the Nobel Prize, Menchú intensified her work on behalf of human rights and indigenous peoples. In 1993, she founded the Rigoberta Menchú Tum Foundation, an organization focused on promoting indigenous rights and empowering these communities, especially in Guatemala and Latin America. The foundation works in education, health, women's rights and the fight against discrimination, as well as strengthening indigenous identity and culture.

Rigoberta Menchú also became involved in her country's politics, to give indigenous peoples a voice and representation in the Guatemalan political system. In 2007 and 2011, she ran for president of Guatemala, although she failed to win. Her participation in the elections was, however, a significant step towards the inclusion of indigenous communities in the country's political process, setting an important precedent and paving the way for future generations of indigenous leaders.

In addition, Menchú has played an active role in defending human rights at the international level, participating in organizations such as the United Nations and collaborating in peace and social justice initiatives in Latin America. Her activism has helped raise awareness about the importance of preserving cultural diversity and respecting the rights of indigenous peoples.

Her legacy lives on in the increased visibility and organization of indigenous peoples in Latin America, in the creation of laws and policies that seek to protect their rights, and in the inspiration, she has provided to young activists around the world.

Her life and work continue to be an example of peaceful struggle, dignity, and perseverance, and her contribution to the indigenous cause remains one of the most important in contemporary history.

21. Benazir Bhutto

She was a prominent Pakistani politician, the first woman to lead a Muslim country in the contemporary era, and one of the most influential figures in Pakistan and the Islamic world. She excelled in the political arena, leading the fight for democracy in her country and promoting reforms aimed at modernization, social welfare, and women's rights in Pakistan. Her legacy is remembered for her courage and determination to challenge authoritarianism, and for having marked a milestone as a symbol of progress and social justice in a political and cultural environment dominated by men.

Born on June 21, 1953, in Karachi, Pakistan, Benazir Bhutto came from a prominent family in Pakistani politics. She was the daughter of Zulfikar Ali Bhutto, founder of the Pakistan People's Party (PPP) and former President and Prime Minister of the country. Benazir received an outstanding education both in Pakistan and abroad, studying at Harvard University and Oxford University, where she became the first Asian woman to chair the prestigious Oxford Union, a debating society. This education abroad gave her an international perspective and a solid preparation in politics and international relations.

In 1977, her father's government was overthrown in a coup led by General Muhammad Zia-ul-Haq, who imposed a military dictatorship in Pakistan. Soon after, Zulfikar Ali Bhutto was arrested and, after a controversial trial, was executed in 1979. These events changed Benazir's life and prompted her to enter politics with a strong commitment to restoring democracy in Pakistan. Following her father's execution, Benazir became leader of the PPP and began an intense campaign against Zia-ul-Haq's military regime, facing arrest, house arrest and several threats to her safety.

In 1988, after General Zia's death in a plane crash, Pakistan held general elections in which the PPP won a majority of seats. At just 35 years old, Benazir Bhutto assumed the position of Prime Minister of Pakistan, becoming the first woman to lead a Muslim country. During her first term, Benazir promoted policies for the improvement of social welfare, focusing on health, education, and women's rights. She implemented vaccination programs and worked to modernize the public health system. In addition, she promoted girls' education and sought to strengthen the role of women in Pakistani society.

However, her term faced strong opposition and conflict, both within and outside her party, and she was dismissed in 1990 on charges of corruption. In 1993, Bhutto was again elected prime minister and continued her efforts to improve the economy and expand social services. Despite her attempts to implement positive changes, her second term was also characterized by corruption in government circles and the growing power of conservatives, which eventually led to her dismissal in 1996.

After being dismissed in 1996, Bhutto went into exile but maintained her influence in Pakistani politics from abroad. She lived between London and Dubai for nearly a decade, as the Pakistani government, under President Pervez Musharraf, was once again under military control. In exile, Bhutto continued to advocate for democracy and seek international support for her cause.

In 2007, after negotiations and an amnesty agreement with President Musharraf, Benazir Bhutto returned to Pakistan to contest parliamentary elections. Her return was marked by great hope among her supporters, who saw her as a figure of change and resistance. However, Pakistan also faced a growing threat from extremism and sectarian violence.

On 27 December 2007, Benazir Bhutto was assassinated in an attack during a rally in Rawalpindi, just a few months after her return to Pakistan. Her death shocked the country and the entire world. Bhutto's assassination left deep sadness among her supporters and created a climate of instability in Pakistan. Investigations into her death point to the involvement of extremist groups, although the true authorship remains a matter of debate.

Benazir Bhutto left a legacy of resilience and struggle for democracy in Pakistan and the Islamic world. Her life and political career were a symbol of the possibility of change and progress in traditionally patriarchal and authoritarian societies. Although her time in power was marked by controversy and conflict, she is remembered as an inspiration to women around the world and to those seeking justice and social reform. Bhutto broke gender barriers in the Muslim world, proving that women could lead and face significant challenges in any field.

Bhutto also inspired future generations of Pakistani activists and politicians, and her fight for democracy in a country with conservative majorities and deep divisions remains a reminder of the challenges and sacrifices faced by those seeking social justice and political change. Her life and death underline the complexity of politics in Pakistan, as well as the strength and courage of a woman who fought to the end for her country and her people.

22. Wangari Maathai

She was a Kenyan environmental activist, politician and academic, known worldwide as the founder of the Green Belt Movement. Maathai was the first African woman to receive the Nobel Peace Prize, in 2004, in recognition of her work in environmental protection, social justice and human rights. Her legacy lies in the creation of an environmental movement in Africa, the promotion of women's rights and social justice, as well as her efforts to raise awareness of the importance of sustainability and ecological development.

Wangari Maathai was born on April 1, 1940, in the Nyeri district of central Kenya, in a farming community. From a young age, she showed great curiosity about the natural world around her and the life of plants and trees, which played a central role in her culture and in the livelihood of her community. At a time when few women had access to education in Kenya, Maathai was an outstanding student, and, thanks to a scholarship, studied Biology at Mount St.

Scholastica University in Kansas, USA. She continued her studies at the University of Pittsburgh, where she earned a master's degree, and subsequently became the first woman in East and Central Africa to earn a PhD, completing it at the University of Nairobi.

In the 1970s, Maathai noticed the environmental deterioration in her country due to massive deforestation, soil erosion and water scarcity, which severely affected the quality of life of rural communities. This situation inspired Maathai to find the Green Belt Movement in 1977, which aimed to promote reforestation and restore the environment, as well as generate employment for rural women. The movement encouraged women to plant trees in their communities to counteract the negative effects of deforestation and provide sustainable sources of water and firewood.

Under her leadership, the Green Belt Movement mobilized thousands of Kenyan women, empowering them and providing them with the training and resources to plant millions of trees in Kenya and elsewhere in Africa. This effort was not only an environmental action, but also a social movement that sought to improve the lives of women and their communities, promoting their economic independence and active role in preserving their environment.

Wangari Maathai's vision went beyond reforestation. She was committed to social justice and human rights, especially those of women and vulnerable communities. Maathai believed that environmental problems and poverty were deeply interconnected, and that environmental damage was also a form of injustice that disproportionately affected the most disadvantaged sectors of society.

Her environmental activism led her to confront the Kenyan government on multiple occasions, especially in its fight against corruption, abuse of power and the grabbing of public lands. In one of her most iconic acts of resistance, Maathai led a campaign to stop the construction of a skyscraper in Nairobi's Uhuru Park, one of the city's few green areas. Thanks to the public pressure generated by her protest, the project was cancelled, marking a major victory for environmental activism in Kenya and demonstrating Maathai's ability to challenge authorities in the name of the common good.

In 2004, Wangari Maathai became the first African woman to receive the Nobel Peace Prize, in recognition of her contribution to sustainable development, democracy and peace. The award was in recognition of her pioneering work in defending the environment as an integral part of social well-being and human rights. In her acceptance speech, Maathai stressed the importance of restoring the balance between humans and nature, and of recognizing that respect for the environment is essential to achieving lasting peace.

This international recognition helped her socialist and social work spread globally, and Maathai became an influential voice in the international environmental movement, speaking at high-profile conferences and events, and promoting her vision of green development around the world.

Wangari Maathai left a profound legacy in the field of environmental sustainability and human rights advocacy in Africa. Her Green Belt Movement has planted more than 51 million trees in Africa and continues to be an active and respected organization in the fight for environmental conservation and restoration. Her innovative approach in combining the themes of environmental justice, women's empowerment, and sustainability has inspired numerous organizations and movements around the world.

Maathai's impact also extends to the empowerment of rural women, as she gave them the tools to take an active role in caring for the environment, demonstrating that they can be leaders in the fight against climate change and environmental degradation. Her life is a testament to how a single person, with determination and commitment, can transform both the social consciousness and the ecological landscape of a nation.

Furthermore, Maathai's work served as a foundation for future generations of environmental leaders and activists in Africa to continue to promote sustainable development and social justice, integrating practices that respect and preserve the natural environment.

Her legacy lives on in every tree planted and every person inspired by her cause represents a vision of a world in which peace, justice and environmental sustainability are deeply connected.

23. Amelia Earhart

She was a pioneer of American aviation and one of the most iconic figures of female aviation in the world. Her legacy lives on as a symbol of courage and perseverance, and her life inspired women of all generations to pursue their dreams no matter the obstacles. Earhart was the first woman to fly solo over the Atlantic Ocean and to set numerous air records, challenging the gender limitations of her time.

Amelia Mary Earhart was born on July 24, 1897, in Atchison, Kansas, United States. From a young age, Amelia displayed an independent and curious personality, excelling in sports such as basketball and traditionally male activities. Earhart began her education at several schools, as her family moved frequently due to her father's financial instability. Her love for aviation began when she attended an air show in 1920 and had her first flight experience in a biplane. From then on, she decided that she would learn to fly and, in 1921, she began taking lessons with Anita "Neta" Snook, a pioneering flight instructor.

Amelia Earhart excelled in the field of aviation, a male-dominated field in which, from its inception, she broke records and demonstrated extraordinary skill. In 1922, just a year after she began flying, she achieved her first female altitude record by reaching 14,000 feet (4,300 meters). In 1928, Earhart became the first woman to cross the Atlantic as a passenger on a flight that received extensive media coverage, although this achievement still did not satisfy her ambition to be the first to cross it alone.

On May 20, 1932, Earhart became the first woman and the second person (after Charles Lindbergh) to cross the Atlantic alone. She flew from Newfoundland to Ireland, a journey of approximately 15 hours. This milestone cemented her fame and respect as an aviator, and she was awarded the Distinguished Flying Cross by the United States Congress. From that point on, Earhart continued to set distance and altitude records and promote commercial aviation, which was just emerging.

Amelia Earhart was a committed activist for women's rights and gender equality. She used her fame to advocate for women's place in traditionally male professions and challenged the gender stereotypes of her time. She was co-founder and first president of the Ninety-Nines, an organization for women pilots created in 1929 that still exists today and seeks to support and make women visible in aviation.

Earhart also wrote articles and gave numerous talks encouraging women to follow their ambitions and not be limited by social conventions. Her motto "women, like men, should attempt the impossible" reflected her belief in equal opportunity and the power of determination.

In 1937, Amelia Earhart embarked on her most ambitious journey: a flight around the world. Together with her navigator, Fred Noonan, Earhart planned to cover approximately 29,000 miles (47,000 kilometers) in 40 days. The flight began on June 1, 1937, in Miami, Florida, and was successful in its early stages. However, on July 2, during one of the last legs of her journey, Earhart's plane disappeared while flying over the Pacific Ocean near Howland Island, with no contact being made with her support crew.

Despite an intensive search involving the United States government, no trace of Earhart or Noonan was ever found, making their disappearance one of aviation's most intriguing mysteries. Since then, various theories have emerged about their fate, including the possibility of a crash landing on an uninhabited island or the plane crashing into the ocean.

Amelia Earhart is remembered as a legend in aviation and as a pioneer of women's rights. Her bravery and achievements challenged the gender expectations of her time and opened doors for women in previously inaccessible areas.

Earhart proved that women could excel in aviation and in any other field, marking a milestone in the history of gender equality.

The figure of Amelia Earhart symbolizes both the pioneering spirit and the passion to break barriers.

24. Virginia Woolf

She was a prominent British writer and essayist, recognized as one of the most influential literary voices of the 20th century. Woolf is famous for her innovative narrative techniques, her introspective style and her deep analysis of human psychology. Her legacy lies both in her contribution to modern literature and in her impact on feminism and the role of women in intellectual culture.

Adeline Virginia Stephen was born on January 25, 1882, in London, England, to an intellectual family. Her father, Sir Leslie Stephen, was a renowned historian and editor, and her mother, Julia Stephen, was a model and nurse. From a young age, Woolf was surrounded by books, and the cultural environment of her home encouraged her to develop an early passion for reading and writing. Despite not receiving a complete formal education, as was common for women of her time, Woolf had access to her father's library and taught herself.

Woolf's life was marked by traumatic experiences, including the death of her mother when she was 13 and that of her father when she was 22. These losses plunged her into bouts of depression that accompanied her throughout her life. The mental disorders she suffered were constant, but they also influenced her perspective and the way she described life and the human mind in her works.

Virginia Woolf excelled in literature, especially in the modern novel, thanks to her experimental style and her ability to capture subjectivity and the flow of thoughts in her characters. With works such as Mrs. Dalloway (1925), To the Lighthouse (1927), and The Waves (1931), Woolf employed an innovative technique known as “interior monologue” or “stream of consciousness,” which explored the inner thoughts and individual perceptions of her characters.

In Mrs. Dalloway, for example, Woolf narrates a single day in the life of Clarissa Dalloway, showing the complexities of her mind and the multiple layers of her social and emotional life. This work, which remains one of her most recognized, highlights her ability to capture the inner lives of her characters in an introspective and poetic way, which marked an evolution in modern narrative.

Con Orlando (1928), Woolf también demostró su audacia al explorar temas de identidad y género. La novela cuenta la historia de un personaje que cambia de sexo y vive a través de siglos, y desafía las convenciones de género y sexualidad. La relación de Woolf con la escritora Vita Sackville-West, que se inspiró en parte de esta obra, influyó en la exploración de estos temas.

In addition to her literary work, Virginia Woolf was one of the most influential feminist thinkers of her time. In her essay A Room of One's Own (1929), Woolf argued for the importance of economic independence and personal space so that women could devote themselves to writing and art. This work introduced the famous phrase: “A woman must have money and a room of her own if she is to write fiction,” a phrase that has become a symbol of the feminist struggle for equal opportunities.

In Three Guineas (1938), Woolf analyzed the role of women in society and how patriarchy affected both men and women, delving into issues of education, economics, and political participation. Her critical view of gender inequalities and her defense of a “society of equals” advanced ideas that would later be fundamental in the feminist movement of the second half of the 20th century.

Throughout her life, Virginia Woolf faced several episodes of depression and nervous breakdowns, exacerbated by stress and personal trauma. Her mental illness manifested itself intermittently, alternating periods of creativity with periods of deep anguish and suicidal thoughts. Despite these challenges, Woolf continued to write and create, finding in literature a refuge and a means to explore her inner experiences.

On March 28, 1941, after a deep crisis of depression, Woolf passed away by drowning herself in the River Ouse, near her home in Sussex. In her farewell note to her husband, Leonard Woolf, she expressed her love and gratitude and spoke of her struggle with mental illness.

Virginia Woolf left a profound legacy in literature and in feminism. Her experimentation with narrative and her ability to portray the inner lives of her characters made her a precursor of modern literature. Her narrative technique and introspective style influenced generations of writers, from William Faulkner to Gabriel García Márquez and Toni Morrison, who have highlighted her ability to explore the complexities of the human mind.

Her feminist vision and writings on women's independence laid the groundwork for what would become literary feminism. A Room of One's Own and Three Guineas remain essential readings in gender and literary studies, and her ideas on the importance of women's economic and emotional freedom continue to inspire activists and writers.

Virginia Woolf is remembered as a revolutionary figure who transformed the way we perceive and express literature, life and identity. Her personal struggle, her courage to challenge conventions and her ability to capture human complexity in words made her a pioneer, whose impact resonates today in art, literature and feminist studies.

25. Aung San Suu Kyi

She is a Burmese political figure and activist who has been widely recognized for her peaceful struggle for democracy and human rights in Myanmar (Burma). Her commitment to non-violence and her leadership in resisting her country's military dictatorship earned her the Nobel Peace Prize in 1991. However, her legacy has become controversial in recent years due to international criticism of her response to the human rights crisis in Myanmar regarding the treatment of the Rohingya minority.

She was born on June 19, 1945, in Rangoon, Burma (now Yangon, Myanmar), the daughter of Aung San, a Burmese independence hero, and Khin Kyi, a diplomat. Her father was assassinated when she was a child, an event that left a deep mark on her life. Educated in Burma and India, Aung San Suu Kyi moved to England to continue her studies at Oxford University, where she studied Philosophy, Politics and

Economics. During her time in the United Kingdom, she met her future husband, the British academic Michael Aris, with whom she had two children.

Years later, in 1988, she returned to Myanmar to care for her ailing mother, but her visit coincided with a period of great political upheaval in the country. The military dictatorship that had controlled Myanmar for decades was facing growing popular opposition, and Aung San Suu Kyi soon became a symbol of peaceful resistance against the regime.

Inspired by figures such as Mahatma Gandhi and Martin Luther King Jr., Aung San Suu Kyi took a nonviolent approach to confronting the military regime. She founded the National League for Democracy (NLD) in 1988 and promoted a peaceful transition to democratic rule. Soon, her leadership and popularity among the Burmese people made her a threat to the military government, and in 1989 she was placed under house arrest, a measure intended to limit her influence in the political opposition.

Aung San Suu Kyi spent nearly 15 of the next 21 years under house arrest but remained a symbolic figure in the fight for human rights and democracy. In 1990, despite her arrest, the NLD won a landslide majority in the general election, but the military junta annulled the results and continued in power. During these years, the international community recognized Aung San Suu Kyi's personal sacrifice and commitment to the cause of freedom, awarding her the 1991 Nobel Peace Prize. In her acceptance speech, her husband and children spoke on her behalf, highlighting her courage and dedication to the principles of peace and justice.

In 2010, Aung San Suu Kyi was finally released from house arrest and returned to lead the NLD. In the 2015 elections, the NLD won a landslide victory, and although the Burmese constitution prevented her from becoming president, the position of State Counsellor was created so that she could rule as the country's de facto leader.

Her rise to power was greeted with great enthusiasm and hope by the Burmese people and the international community, who saw in her a symbol of justice and democracy. However, a human rights crisis erupted in 2017 when the Burmese military carried out a violent crackdown on the Rohingya Muslim minority, resulting in thousands of deaths and the displacement of more than 700,000 people to Bangladesh. Aung San Suu Kyi's response was perceived as passive and, in some cases, defensive towards the military's actions, leading to widespread criticism and disillusionment from those who once regarded her as a human rights icon.

Aung San Suu Kyi's legacy is complex. On the one hand, her leadership in the fight for democracy inspired generations of people in Myanmar and around the world. Her personal sacrifice and dedication to a just cause, even while under house arrest, made her a respected and admired figure. As one of the few women in history to have led a peaceful resistance movement against a military regime, Aung San Suu Kyi remains a symbol of courage and persistence for many.

However, her legacy has been tarnished by her response to the Rohingya crisis, leading to some of her international awards and honours being questioned and even withdrawn. This situation has raised debates about the complexity of political leadership in conflict situations and the limitations of acting in a system still largely controlled by the military.

Aung San Suu Kyi's story continues to be a source of reflection on the moral and ethical dilemmas that accompany power, as well as the difficulties of implementing genuine change in challenging political contexts.

———†———

Other books by the author Phillips Tahuer that you will find on this platform:

• The greatest conspiracy theories

• Great robberies in history

• Famous murderers - the perverse side of the mind-

• Lives in captivity - Stories of real kidnappings-

• Agents, informants and traitors - the world of espionage-

• Pirates of the 21st century

• Tragic loves

• 30 curiosities of World War II

• Dark experiments on humans

• Real-life heroes

• Powerful men in modern history

• Valentine's Day stories

• Lessons in practical psychology

• Deserters

• Attacks and assassinations

• Aviation aces